THE POETRY OF THALLIUM

The Poetry of Thallium

Walter the Educator

Silent King Books

SILENT KING BOOKS

SKB

Copyright © 2024 by Walter the Educator

All rights reserved. No part of this book may be reproduced in any manner whatsoever without written permission except in the case of brief quotations embodied in critical articles and reviews.

First Printing, 2024

Disclaimer
This book is a literary work; poems are not about specific persons, locations, situations, and/or circumstances unless mentioned in a historical context. This book is for entertainment and informational purposes only. The author and publisher offer this information without warranties expressed or implied. No matter the grounds, neither the author nor the publisher will be accountable for any losses, injuries, or other damages caused by the reader's use of this book. The use of this book acknowledges an understanding and acceptance of this disclaimer.

"Earning a degree in chemistry changed my life!"
- Walter the Educator

dedicated to all the chemistry lovers, like myself, across the world

THALLIUM

Thallium's, a silent dance of atoms plays,

THALLIUM

Where electrons whirl in mystical arrays.

THALLIUM

A metal with a luster, both eerie and profound,

THALLIUM

In the periodic table, its essence is found.

THALLIUM

From ancient lore to modern labs, its story weaves,

THALLIUM

A tale of secrets hidden beneath autumn leaves.

THALLIUM

With atomic number eighty-one, it holds its place,

THALLIUM

In the elements' symphony, it finds its grace.

THALLIUM

Beneath a cloak of mystery, Thallium resides,

THALLIUM

In soil and water, where its presence abides.

THALLIUM

From mining depths to the depths of our minds,

THALLIUM

Its allure beckons, where the curious binds.

THALLIUM

In compounds, it ventures, a chameleon's guise,

THALLIUM

With oxides and halides, it mesmerizes.

THALLIUM

In salts and solutions, its toxicity reigns,

THALLIUM

A cautionary tale, where danger feigns.

THALLIUM

Yet amidst its shadows, a glimmer of light,

THALLIUM

In medicine's grasp, it fights the blight.

THALLIUM

From poison to potion, its journey unfolds,

THALLIUM

A healer's touch, where alchemy molds.

THALLIUM

In atomic clocks, its ticks measure time's span,

THALLIUM

A heartbeat of science, where precision's plan.

THALLIUM

In spectral lines, it whispers cosmic tales,

THALLIUM

Of stars and galaxies, where wonder sails.

THALLIUM

Its isotopes dance in nuclear decay,

THALLIUM

A cosmic waltz, where particles sway.

THALLIUM

In fission's fury or fusion's embrace,

THALLIUM

Its energy pulses through time and space.

THALLIUM

In the artist's palette, its colors bloom,

THALLIUM

A painter's dream, where hues consume.

THALLIUM

In stained glass windows, it casts its spell,

THALLIUM

A kaleidoscope of stories to tell.

THALLIUM

In the poet's ink, its verses flow,

THALLIUM

A lyric ode, where meanings grow.

THALLIUM

In every stanza, a spark ignites,

THALLIUM

A tribute to Thallium's mystic lights.

THALLIUM

So let us marvel at this element's tale,

THALLIUM

Where science and wonder forever prevail.

THALLIUM

In Thallium's domain, a universe untold,

THALLIUM

A symphony of atoms, infinitely bold.

THALLIUM

ABOUT THE CREATOR

Walter the Educator is one of the pseudonyms for Walter Anderson. Formally educated in Chemistry, Business, and Education, he is an educator, an author, a diverse entrepreneur, and he is the son of a disabled war veteran. "Walter the Educator" shares his time between educating and creating. He holds interests and owns several creative projects that entertain, enlighten, enhance, and educate, hoping to inspire and motivate you.

Follow, find new works, and stay up to date with Walter the Educator™ at WaltertheEducator.com

www.ingramcontent.com/pod-product-compliance
Lightning Source LLC
LaVergne TN
LVHW010412070526
838199LV00064B/5277

* 9 7 9 8 8 6 9 2 8 6 2 0 8 *